Year 7, Practice Book 3

KU-346-211

NEW MATHS FRAMEWORKING

Matches the revised KS3 Framework

Kevin Evans, Keith Gordon, Trevor Senior, Brian Speed

William Collins' dream of knowledge for all began with the publication of his first book in 1819. A self-educated mill worker, he not only enriched millions of lives, but also founded a flourishing publishing house. Today, staying true to this spirit, Collins books are packed with inspiration, innovation and practical expertise. They place you at the centre of a world of possibility and give you exactly what you need to explore it.

Collins. Freedom to teach.

Published by Collins
An imprint of HarperCollins*Publishers*
77–85 Fulham Palace Road
Hammersmith
London
W6 8JB

Browse the complete Collins catalogue at
www.collinseducation.com

10 9 8 7 6 5 4

ISBN 978-0-00-726793-4

British Library Cataloguing in Publication Data
A Catalogue record for this publication is available from the British Library.
Commissioned by Melanie Hoffman and Katie Sergeant
Project management by Priya Govindan
Covers management by Laura Deacon
Edited by Karen Westall
Proofread by Amanda Dickson
Design and typesetting by Newgen Imaging
Design concept by Jordan Publishing Design
Covers by Oculus Design and Communications
Illustrations by Derek Lee and Newgen Imaging
Printed and bound by Printing Express, Hong Kong
Production by Simon Moore

Mixed Sources
Product group from well-managed forests and other controlled sources
www.fsc.org Cert no. SW-COC-1806
© 1996 Forest Stewardship Council

FSC is a non-profit international organisation established to promote the responsible management of the world's forests. Products carrying the FSC label are independently certified to assure consumers that they come from forests that are managed to meet the social, economic and ecological needs of present and future generations.

Find out more about HarperCollins and the environment at
www.harpercollins.co.uk/green

Welcome to *New Maths Frameworking*!

New Maths Frameworking Year 7 Practice Book 3 has hundreds of levelled questions to help you practise Maths at Levels 5-6. The questions correspond to topics covered in Year 7 Pupil Book 3 giving you lots of extra practice.

These are the key features:

- **Colour-coded National Curriculum levels** for all the questions show you what level you are working at so you can easily track your progress and see how to get to the next level.

- **Functional Maths** is all about how people use Maths in everyday life. Look out for the Functional Maths icon **FM** which shows you when you are practising your Functional Maths skills.

Contents

CHAPTER 1 Algebra 1

1A Sequences and rules

1 Use each of the following term-to-term rules with 1st term 2. Create the sequences with five terms in each.

a Add 3   **b** Add 6   **c** Treble
d Multiply by 5   **e** Add 100   **f** Multiply by 10

2 Write the next two terms in each sequence. Describe the term-to-term rule you have used.

a 1, 3, 5, 7, …   **b** 20, 30, 40, 50, …   **c** 5, 13, 21, 29, …
d 5, 10, 15, 20, …   **e** 6, 13, 20, 27, …   **f** 10, 110, 210, 310, …

3 Find two terms between each pair of numbers to form a sequence. Describe the term-to-term rule you have used.

a 2, … , … , 8   **b** 5, … , … , 14   **c** 6, … , … , 18
d 3, … , … , 24   **e** 1000, … , … , 1   **f** 16, … , … , 2

4 **a** Add together any two sequences from question 2, term-by-term.
     b Write down the 1st term and term-to-term rule.
     c You could answer part **b** without adding the sequences. How?

1B Finding missing terms

1 In each of the following sequences, find the 5th and 50th terms.

a 1, 5, 9, 13, …   **b** 3, 5, 7, 9, …   **c** 4, 12, 20, 28, …
d 5, 15, 25, 35, …   **e** 2, 8, 14, 20, …   **f** 10, 30, 50, 70, …
g 2, 5, 8, 11, …   **h** 0, 5, 10, 15, …   **i** 4, 11, 18, 25, …

2 In each of the following sequences, find the missing terms and the 30th term.

Term	1st	2nd	3rd	4th	5th	6th	7th	8th	…	30th
Sequence A	—	—	—	13	16	19	22	—	…	—
Sequence B	—	9	16	—	30	37	—	—	…	—
Sequence C	—	—	25	—	45	—	65	—	…	—
Sequence D	—	11	—	19	—	27	—	—	…	—

6

1C Finding the general term (*n*th term)

1 For each of the sequences whose *n*th term is given below, find the following.

i The first three terms   **ii** The 100th term

a $3n - 1$ **b** $5n + 2$ **c** $6n - 5$
d $10n - 1$ **e** $3n + 8$ **f** $\frac{1}{2}n + 1\frac{1}{2}$

2 Find the *n*th term for each of the following patterns. Use this generalisation to find the 40th term in each pattern.

a

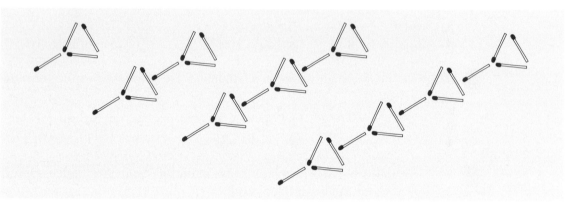

b

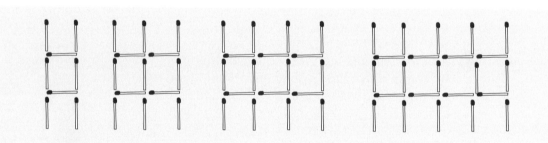

3 Find the *n*th term of each of the following sequences.

a 4, 10, 16, 22, 28, …
b 8, 11, 14, 17, 20, …
c 9, 15, 21, 27, 33, …
d 4, 7, 10, 13, 16, …
e 13, 20, 27, 34, 41, …

1D Functions and mappings

1 Express each of these simple functions in words.

a →[   ]→   **b** →[   ]→   **c** →[   ]→

input	output
3	9
4	12
5	15
6	18

input	output
3	10
4	11
5	12
6	13

input	output
8	2
12	3
16	4
20	5

4

2

2 Give the missing values in these double function machines.

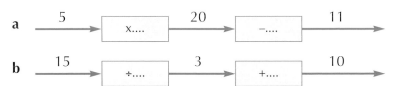

a 5 →[x....]→ 20 →[–....]→ 11

b 15 →[÷....]→ 3 →[+....]→ 10

3 Each of the following functions is made up from two operations as above. Find each of the combined functions.

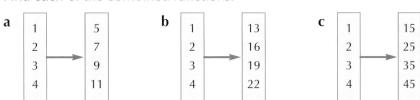

a
1		5
2	→	7
3		9
4		11

b
1		13
2	→	16
3		19
4		22

c
1		15
2	→	25
3		35
4		45

4 Work backwards from the output to find the input to each of the following functions.

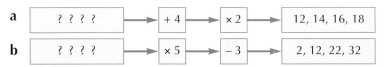

a [? ? ? ?] → [+ 4] → [× 2] → 12, 14, 16, 18

b [? ? ? ?] → [× 5] → [– 3] → 2, 12, 22, 32

Practice

1E Using letter symbols to represent functions

1 Write each of the following rules in symbolic form, e.g. $x \to x + 4$.

   a Subtract 5   **b** Treble   **c** Add 9   **d** Divide by 3

2 Draw mapping diagrams to illustrate each of the following functions.

   a $x \to x + 25$ **b** $x \to 3x$ **c** $x \to 3x - 1$ **d** $x \to 4x + 4$

3 Express each of the following functions in symbolic form, as in question 1.

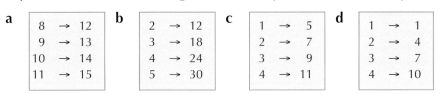

a
8	→	12
9	→	13
10	→	14
11	→	15

b
2	→	12
3	→	18
4	→	24
5	→	30

c
1	→	5
2	→	7
3	→	9
4	→	11

d
1	→	1
2	→	4
3	→	7
4	→	10

4 Find the inverse of each function in question 3.

5 Put the same number through each of these function machines.

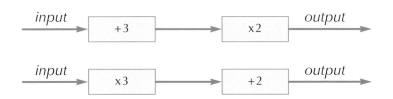

input →[+3]→[x2]→ output

input →[x3]→[+2]→ output

Repeat with other numbers.
Can you find an input that gives the same output for both function machines?

6 The diagram shows a number machine.

| INPUT | → | × 4 | → | − 24 | → | OUTPUT |

a Write down the output when the input is 5.
b Find a number to input, so that the output is the same as the input.

CHAPTER 2 Number 1

Practice

2A Decimals

1 Without using a calculator, work out each of the following.

a	1.3 × 10	b	0.06 × 10	c	0.058 × 100
d	0.74 × 100	e	0.04 × 1000	f	0.8 ÷ 10
g	1.09 ÷ 10	h	0.62 ÷ 100	i	2.9 ÷ 100
j	4.97 × 10	k	0.08 ÷ 100	l	1.412 × 100

2 Fill in each missing operation.

a | 0.6 | → | | → | 0.06 | b | 0.65 | → | | → | 65 |

c | 432 | → | | → | 4.32 | d | 0.265 | → | | → | 265 |

3 Find each missing number.

a $0.6 \times 100 = \boxed{}$ b $6 \div \boxed{} = 0.6$

c $0.6 \times \boxed{} = 600$ d $\boxed{} \div 100 = 0.06$

e $600 \times \boxed{} = 60\,000$ f $0.6 \div 10 = \boxed{}$

4 Copy and complete each of the following.

a $10^8 = \dots \times \dots \times \dots \times \dots \times \dots \times \dots \times \dots \times \dots = \dots\dots$
b $10^{10} = \dots\dots\dots\dots\dots\dots\dots\dots\dots\dots\dots\dots\dots = \dots\dots$

5 Write down the answer to each of the following.

a	5.4×10^3	b	0.6×10^4	c	0.49×10^3
d	$0.9 \div 10^3$	e	$800 \div 10^3$	f	$0.04 \div 10^3$

Practice

2B Ordering decimals

1. **a** Copy the table on page 18 of Pupil Book 3 (but not the numbers). Write the following numbers in the table, getting each digit in the appropriate column.

   0.02903, 0.293, 2.93, 0.2093, 2.093, 0.29, 0.0029

   b Use your answer to part **a** to write the numbers in order from smallest to largest.

2. Write each set of numbers in order from smallest to largest.

   a 3.015, 3.1050, 3.0015, 3.15, 3.051
   b 0.0132, 0.031, 0.302, 0.123, 0.0032

3. Working in kilograms, put the weights below in order, from smallest to largest.

| 0.785 kg | 0.67 kg | 3.2 kg | 0.652 kg | 0.8 kg |

4. Put these amounts of money in increasing order.
   £1.20   32p   £0.28   23p   £0.63

5. Put these lengths in decreasing order.
   57 cm, 2.05 m, 0.06 m, 123 cm, 0.9 m

6. Put the correct sign, > or <, between each pair of numbers.

   a 3.62 … 3.26   **b** 0.07 … 0.073   **c** £0.09 … 10p

Practice

2C Directed numbers

1. Work out the answer to each of these.

   a −5 + 8   **b** 2 − 7   **c** −4 − 9   **d** 2 − 6 − 3
   e −5 + 6 − 4   **f** −3 + 8 − 2   **g** 5 − +7   **h** −7 + −3
   i 4 − −6   **j** − +7 − 4 + 6   **k** 2 − −5 − −8   **l** −4 + −6 − +3

2. Find the missing numbers to make each of these true.

   a −5 + ☐ = −3   **b** ☐ − 3 = −7

   c ☐ − −3 = 0   **d** −2 − ☐ = 6

3 In a magic square, the numbers in any row, column or diagonal add up to give the same answer. Could this be a magic square? Give a reason for your answer.

−3		−7
		4
		−5

4 Work out the answer to each of these.

a −5 × −2   b +6 × −4   c −3 × +8

d +9 × −6   e −7 × +7   f −6 × −8

g +5 × −3 × +4   h −3 × −2 × −1

5 Work out the answer to each of these.

a −16 ÷ −2   b +12 ÷ −4   c −18 ÷ +3

d +30 ÷ −5   e −36 ÷ +6   f −100 ÷ −20

g −8 × −2 ÷ +4   h +12 ÷ −4 × 3

6 Copy and complete the multiplication grid.

×	−4			2
	−20		−25	
−3				
		−48	40	
4				

2D Estimates

1 Explain why these calculations must be wrong.

a 53 × 21 = 1111   b 58 × 34 = 2972   c 904 ÷ 14 = 36

2 Round each of the following numbers to one decimal place.

a 0.84   b 0.45   c 5.21   d 3.87   e 2.06

f 2.03   g 0.129   h 6.048   i 5.96   j 3.98

3 Work out each of these.

a 50 × 0.3   b 20 × 0.7   c 80 × 0.6

d 90 × 0.3   e 0.4 × 70   f 130 × 0.3

4 Estimate the answer to each of these problems.

a 6832 − 496   b 28 × 123   c 521 ÷ 18

d 770 × 770   e $\dfrac{58.9 + 36.4}{22.5}$

5 **a** Football socks cost £3.71 a pair. Without calculating the answer, could Ian buy five pairs using a £20 note? Explain your answer clearly.

b Shoelaces cost 68p a pair. The shopkeeper charges Ian £3.25 for five pairs. Without finding the correct answer, explain why this is wrong.

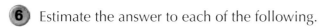

6 Estimate the answer to each of the following.

    a   58 × 0.26   **b**   24 × 0.71   **c**   68 × 0.29

    d   93 × 0.14   **e**   0.36 × 36   **f**   109 × 0.94

7 62 ÷ 0.39 can be approximated as 60 ÷ 0.4 = 600 ÷ 4 = 150. Estimate the answer to each of the following divisions.

    a   83 ÷ 0.23   **b**   48 ÷ 0.07   **c**   37 ÷ 0.44

    d   97 ÷ 0.14   **e**   136 ÷ 0.68   **f**   244 ÷ 0.55

Practice

2E Column method for addition, subtraction and multiplication

1 By means of a drawing, show how you would use a number line to work out the answer to each of these.

    a   4.7 + 3.8   **b**   9.3 – 3.16

2 Repeat the calculations in question 1 using the column method. Show all your working.

3 Use the column method to work out each of these additions.

    a   19.8 + 23.1   **b**   28.4 + 6.92

    c   4.32 + 9.81   **d**   21.38 + 106.26 + 9.34

4 Use the column method to work out each of these subtractions.

    a   17.6 – 9.4   **b**   28.6 – 12.93

    c   5.9 + 8.32 – 4.71   **d**   11.91 + 86.23 – 53.8

 5 Work out the cost of a CD cleaning kit at £7.23, a CD carrying case at £5.69 and a sheet of CD labels at 84p.

 6 A radio costs £26.48 and a CD player costs £52.13. How much dearer is the CD player?

7 Use the column method to work out each of the following.

    a   3.2 × 8   **b**   9.1 × 6   **c**   5.7 × 6   **d**   3.84 × 4

    e   7.29 × 3   **f**   5.07 × 7

 8 Calculate the total cost of nine ties at £3.64 each.

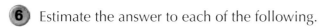

Practice

2F Solving problems

FM **1** Seven cans of beans weigh 1750 g. How much do nine cans weigh?

FM **2** An egg box containing two eggs weighs 140 g. The same box containing three eggs weighs 195 g. What does a box containing six eggs weigh?

3 If 26 × 152 = 3952, write down, without calculating, the value of the following.

    a   2.6 × 152   **b**   2.6 × 15.2   **c**   260 × 1520

4 Find four consecutive even numbers that add up to 60.

5 To make a number chain, start with any number.
When the number is even, divide it by 2 and subtract 1.
When the number is odd, subtract 1 and double the answer.
Investigate which numbers give the longest chains.

CHAPTER 3   Geometry and Measures 1

Practice

3A Length, perimeter and area

1 Copy these puzzle pieces onto 1 cm squared paper. Find the perimeter and area of each piece.

    a 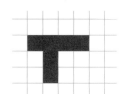   **b**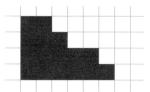

2 Estimate the area of each of these shapes. Each square on the grid represents one square centimetre.

    a 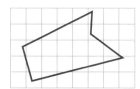   **b**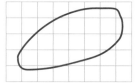

3 **a** Find the perimeter of each rectangle.
   b Find the area of each rectangle.

i 7 m, 16 m

ii 22 mm, 9 mm

iii 6 cm, 6 cm

iv 3 mm, 28 mm

4 Use centimetre squared paper to draw three different rectangles, each with an area of 24 cm².

5 A swimming pool is 8 m wide and 30 m long.

a Find the perimeter of the pool.

b Find the area of the pool.

c Emma wants to swim 1 kilometre.
   i How many widths does she need to swim?
   ii How many lengths does she need to swim?

d The floor of the pool is covered with square tiles of side 50 cm. How many tiles cover the floor?

6 **a** Find the perimeter of each of the following compound shapes.
   b Find the area of each of them.

i 5 mm, 20 mm, 20 mm, 10 mm

ii 12 cm, 9 cm, 5 cm, 3 cm

7 Calculate the area of brickwork on the front of this house.

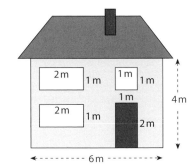

6

1 The shapes below are cut from waterproof material to make tents. Calculate the area of each shape.

a

5 m
2 m

b
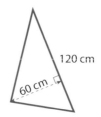
120 cm
60 cm

c
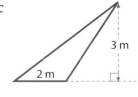
3 m
2 m

d
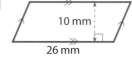
10 mm
26 mm

e
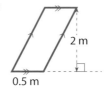
2 m
0.5 m

f
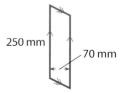
250 mm
70 mm

g
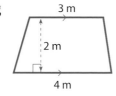
3 m
2 m
4 m

h
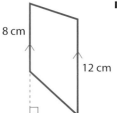
8 cm
12 cm
6 cm

i
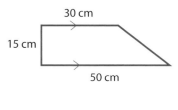
30 cm
15 cm
50 cm

2 The area of each shape is given. Calculate the length of the base.

a
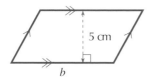
5 cm
b

Area = 60 cm²

b
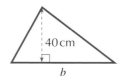
40 cm
b

Area = 200 cm²

3 These shapes are cut from a rectangle of material.

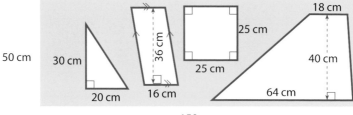

50 cm
30 cm
20 cm
36 cm
16 cm
25 cm
25 cm
18 cm
40 cm
64 cm
150 cm

a Calculate the total area of the shapes.
b Calculate the area of material left over.

3C 3-D shapes

1 Below are the nets of two solids. Describe each solid.

a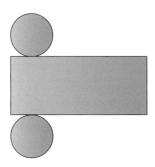

b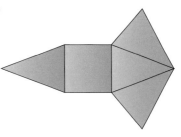

2 On squared paper, draw an accurate net for each model storage box.

a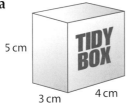

b

3 Draw accurately on an isometric grid each of the following solids.

a

b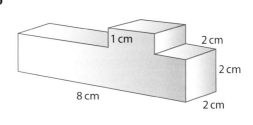

4 Use an isometric grid to draw solids made up of the following.

  a 4 cubes   **b** 10 cubes

3D Surface area and volume of cuboids

1 Find: **a** the surface area and **b** the volume of each of the following cuboids.

i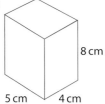

ii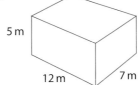

iii

2 Find: **a** the surface area and **b** the volume of each of the cubes with the following edge length.

i 3 cm   **ii** 6 cm   **iii** 12 cm

3 Calculate the total surface area of the inside and outside of this wooden box, including the lid (ignore the thickness of wood).

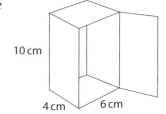

10 cm

4 cm   6 cm

4 How many cubic dice, each of side 2 cm, can be packed into the box in question 3.

5 Use an isometric grid to draw different shapes using six 1 cm cubes. Find the shapes with the greatest surface area.

CHAPTER **4** Number **2**

Practice

4A Fractions

1 Copy and complete the following equivalent fraction series.

$$\frac{2}{3} = \frac{4}{\square} = \frac{\square}{9} = \frac{10}{\square} = \frac{\square}{30} = \frac{60}{\square}$$

2 Find the missing number in each of these equivalent fractions.

a $\dfrac{3}{4} = \dfrac{\square}{16}$ **b** $\dfrac{4}{7} = \dfrac{20}{\square}$ **c** $\dfrac{7}{10} = \dfrac{\square}{70}$ **d** $\dfrac{14}{3} = \dfrac{\square}{15}$

3 Cancel each of these fractions to its simplest form.

a $\frac{5}{20}$ **b** $\frac{10}{12}$ **c** $\frac{21}{35}$ **d** $\frac{16}{20}$ **e** $\frac{12}{18}$ **f** $\frac{30}{25}$
g $\frac{36}{27}$ **h** $\frac{100}{70}$ **i** $\frac{120}{64}$

4 The diagram shows the dial of a combination safe lock.

What fraction of a turn takes you from:

a A to C clockwise?   **b** A to C anticlockwise?
c H to E clockwise?   **d** J to D anticlockwise?
e F to B anticlockwise?   **f** C to H clockwise?

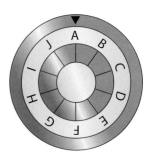

5 Use the information in question 4 to write each of these as a fraction of a whole turn.

    a A to I to D clockwise   **b** F to J to A anticlockwise

    c H to D to A to F clockwise   **d** E to D to C to B anticlockwise

6 **a** 1 day is 24 hours. What fraction of a day is 18 hours?

    b 1 kilobyte is 1000 bytes. What fraction of a kilobyte is 450 bytes?

7 Write each of these fractions as a mixed number. Cancel down if appropriate.

    a Fourteen thirds   **b** $\frac{19}{2}$ **c** $\frac{26}{4}$ **d** $\frac{59}{10}$ **e** $\frac{80}{12}$ **f** $\frac{72}{7}$

Practice

4B Fractions and decimals

1 Convert each of the following decimals to a fraction.

    a 0.7   **b** 0.6   **c** 0.45   **d** 0.16   **e** 0.08   **f** 0.39

2 Convert each of the following fractions to a decimal.

    a $\frac{7}{10}$ **b** $\frac{19}{50}$ **c** $\frac{6}{25}$ **d** $\frac{17}{20}$ **e** $\frac{31}{25}$ **f** $\frac{9}{100}$

3 Covert each of these top–heavy (improper) fractions to a decimal.

    a $\frac{5}{4}$ **b** $\frac{13}{10}$ **c** $\frac{52}{20}$ **d** $\frac{45}{25}$ **e** $\frac{130}{50}$ **f** $\frac{75}{20}$

4 Put the correct sign, < or >, between each pair of fractions.
(Hint: Convert to fractions out of 100.)

    a $\frac{37}{100} \ldots \frac{17}{50}$ **b** $\frac{3}{10} \ldots \frac{11}{50}$ **c** $\frac{13}{20} \ldots \frac{17}{25}$

5 Put these fractions in order of size, smallest first. Use the hint in question 4.

    a $\frac{7}{20}, \frac{3}{10}, \frac{22}{50}$ **b** $\frac{18}{25}, \frac{3}{4}, \frac{71}{100}$ **c** $1\frac{7}{20}, 1\frac{1}{4}, 1\frac{13}{50}$

6 Convert each of these decimals to a mixed number.

    a 1.75   **b** 2.1   **c** 3.4   **d** 4.625

Practice

4C Adding and subtracting fractions

1 **a** Add the following fractions.

      i $\frac{3}{8} + \frac{1}{4}$ **ii** $1\frac{1}{2} + \frac{5}{8}$ **iii** $2\frac{1}{4} + 1\frac{3}{8}$ **iv** $1\frac{1}{4} + \frac{7}{8} + 1\frac{1}{8}$

b Subtract the following fractions.

    i $\frac{7}{8} - \frac{1}{4}$ **ii** $1\frac{1}{2} - \frac{5}{8}$ **iii** $2\frac{3}{8} - \frac{3}{4}$ **iv** $3\frac{5}{8} - 1\frac{3}{4}$

2 **a** Add the following fractions. Convert each answer to a mixed number and cancel to lowest terms where appropriate.

    i $\frac{3}{7} + \frac{2}{7}$ **ii** $\frac{5}{8} + \frac{7}{8}$ **iii** $\frac{2}{11} + \frac{6}{11} + \frac{10}{11}$ **iv** $1\frac{3}{5} + \frac{4}{5}$

 b Subtract the following fractions.

    i $\frac{9}{10} - \frac{7}{10}$ **ii** $\frac{3}{8} - \frac{1}{8}$ **iii** $\frac{13}{16} - \frac{7}{16}$ **iv** $\frac{11}{8} - \frac{5}{8}$

3 Convert each pair of fractions to their equivalent fractions with a common denominator. Then work out the answer, cancelling down and writing as a mixed number where appropriate.

a $\frac{2}{3} + \frac{1}{6}$	**b** $\frac{1}{4} + \frac{2}{5}$	**c** $\frac{1}{3} + \frac{2}{5}$	**d** $\frac{1}{2} + \frac{1}{3} + \frac{1}{4}$
e $\frac{5}{8} - \frac{1}{4}$	**f** $\frac{9}{10} - \frac{1}{2}$	**g** $\frac{3}{5} - \frac{1}{3}$	**h** $\frac{2}{3} + \frac{5}{6} - \frac{1}{9}$

4 Work along each chain of calculations. Write down your answer to each step.

a $\frac{7}{10}$ → $+\frac{3}{10}$ =   → $-\frac{4}{10}$ =   → $+\frac{1}{10}$ =   → $-\frac{6}{10}$ =

b $1\frac{3}{8}$ → $-\frac{5}{8}$ =   → $+\frac{3}{4}$ =   → $-\frac{3}{8}$ =

c $\frac{3}{8}$ → $+1\frac{3}{4}$ =   → $+\frac{5}{8}$ =   → $-1\frac{1}{2}$ =   → $+2\frac{1}{8}$ =   → $-\frac{7}{8}$ =

Practice

4D Calculations and equivalences

1 Calculate the following.

a 10% of £560	**b** 80% of 160 m	**c** 15% of 90 kg
d 40% of 35 kg	**e** 35% of 90p	**f** 85% of 3000 t

2 Work out the following.

  a A quarter of thirty-two   **b** A seventh of forty-nine
  c A tenth of two hundred and fifty

3 Work out each of these, cancelling down and writing as a mixed number where appropriate.

  a $\frac{1}{4}$ of £60   **b** $\frac{3}{5}$ of 30 kg   **c** $\frac{4}{7}$ of 56 m   **d** $\frac{8}{9}$ of 63p

4 Work out each of these, cancelling down where appropriate. Write the answer as a mixed number where appropriate.

a $6 \times \frac{2}{5}$	**b** $9 \times \frac{5}{6}$	**c** $2 \times \frac{6}{7}$	**d** $8 \times \frac{7}{10}$
e $\frac{3}{4} \div 2$	**f** $\frac{6}{7} \div 3$	**g** $\frac{8}{9} \div 6$	**h** $\frac{4}{3} \div 5$

5 Work out the equivalent percentage and fraction to each decimal.

    a   0.7   **b**   0.55   **c**   0.92   **d**   0.02   **e**   0.42

6 Work out the equivalent decimal and fraction to each percentage.

    a   30%   **b**   16%   **c**   95%   **d**   5%   **e**   73%

7 Work out the equivalent percentage and decimal to each fraction.

    a $\frac{4}{10}$ **b** $\frac{17}{20}$ **c** $\frac{9}{25}$ **d** $\frac{37}{50}$ **e** $\frac{4}{5}$ **f** $\frac{5}{8}$

8 A photocopier is set to enlarge the lengths of images by 20%. What will be the new measurements when these objects are enlarged?

    a   20 cm   **b**   **c**   8 cm

                                  150 mm

9 An improved assembly line reduces the time to assemble gamepads by 10%. What will the following assembly times be reduced to?

    a   120 seconds   **b**   95 seconds   **c**   14 seconds

10 Find the pairs of equivalent numbers. Write your answers like this: **m = n**.

    a   40%

    b $\frac{3}{8}$

    c   0.3

    d   0.65

    e   18%

    f $\frac{2}{5}$

    g   30%

    h   0.375

    i $\frac{13}{20}$

    j $\frac{9}{50}$

Practice

4E Solving problems

1 What number is halfway between the numbers shown on each of the following scales?

    a $\frac{1}{8}$ $\frac{5}{8}$ **b** $\frac{5}{12}$ $\frac{11}{12}$ **c** $1\frac{3}{8}$ $2\frac{1}{8}$

2 Which is greater?

    a $\frac{3}{5}$ of 55 or $\frac{3}{4}$ of 48   **b** $\frac{5}{8}$ of 56 or $\frac{4}{7}$ of 49   **c** $\frac{5}{6}$ of 33 or $\frac{7}{10}$ of 35

3 Find one fraction that is between each pair of fractions.

    a $\frac{4}{25}, \frac{1}{5}$ **b** $\frac{11}{20}, \frac{6}{10}$ **c** $\frac{33}{50}, \frac{3}{4}$

(4) A bag contains 240 jelly beans. $\frac{1}{6}$ are red, $\frac{3}{8}$ are yellow, $\frac{2}{5}$ are blue and the rest are green. How many of each colour are there? What fraction is green?

 (5) Which of these sales is the best deal?

a **SALE 35% off**   b **SALE ⅖ off**   c **SALE ¼ off**

 (6) To calculate the tax paid on an annual salary, follow these steps:

▼ Subtract £4500 from the salary. This amount is not taxed.
▼ Calculate 10% of the next £1500 of the salary.
▼ Calculate 25% of the remaining salary.
▼ Add together the results of the last two steps.

Calculate the tax due for each of these salaries.

(a) £6000   (b) £26 000   (c) £7000

CHAPTER 5 Statistics 1

Practice

5A Mode, median and range

(1) Find the mode, median and range for each of the following sets of data.

a   5, 1, 7, 7, 3, 1, 5, 2, 6, 5, 3, 2, 4, 9, 8
b   2, 7, 3, 12, 9, 15, 3
c   31, 19, 17, 28, 40, 30, 42, 7, 17
d   103, 105, 101, 107, 101, 103, 101, 107, 105
e   20, 70, 30, 25, 90, 35, 60, 60, 15
f   3.1, 5.6, 2.9, 4.8, 3.6, 4.9, 6.3, 3.7, 5.6
g   0.7, 0.6, 0.5, 0.9, 0.1, 0.5, 0.8

(2) Find the mode, median and range for each of the following sets of data.

a   200 mm, 150 mm, 600 mm, 300 mm, 450 mm, 200 mm, 500 mm, 250 mm, 550 mm
b   £3.20   £2.90   £3.10   £3   £3.15   £2.90
c   58 g, 48 g, 48 g, 52 g, 50 g, 51 g, 50 g, 56 g, 48 g, 58 g, 52 g, 56 g

3 Seventeen cars of the same year and model were tested for fuel efficiency. The table shows the miles per gallon for the cars.

Miles per gallon	36	37	38	39	40
Number of cars (frequency)	3	4	7	2	1

    a Calculate the mode.
    b Calculate the median.
    c Calculate the range.

4 The table shows the sizes of breaks during a snooker club tournament.

Break	Tally	Frequency
1 – 10	⅂⅂⅂ ⅂⅂⅂ ///	
11 – 20	⅂⅂⅂ ⅂⅂⅂ ⅂⅂⅂ ⅂⅂⅂ ⅂⅂⅂	
21 – 30	⅂⅂⅂ /	
31 – 40	////	
41 – 50	//	

    a Fill in the frequency column.
    b What is the maximum possible range?
    c In which class is the median?
    d Draw a bar chart to illustrate the data.

5   **a** Write down five numbers with a mode of 3 and a median of 4.
    b Write down five numbers with a mode of 10, median of 7 and range of 5.

Practice

5B The mean

1 Find the mean for each of the following sets of data.

    a 3, 15, 7, 10, 23, 4, 9, 1
    b 1.9, 4.2, 2.6, 3, 5.1, 1.8

2 Find the mean for each of the following sets of data, giving your answer to one decimal place.

    a 3, 8, 4, 7, 9, 2, 4
    b 53, 94, 21, 64, 100, 17, 35, 88, 42

3 The lap times, in seconds, for eight model racing cars are shown below.
      17, 20, 15, 16, 20, 15, 17, 20

    a Find the mean lap time.
    b Find the median lap time.
    c Find the modal lap time.
    d Which average do you think is the worst one to use? Explain your answer.
    e Why are so few lap times greater than the mean?

The frequency table shows the numbers of low-value stamps sold at a post office one morning.

Stamp value (p)	1	2	3	4	5
Frequency	25	7	1	2	15

Calculate the mean stamp value.

5 Paddy's chickens laid the following quantities of eggs during one week.

Monday	Tuesday	Wednesday	Thursday
7	6	8	5

a Find the mean number of eggs laid from Monday to Thursday.

The chickens laid more eggs on Friday. The mean for all five days is 7.

b How many eggs did the chickens lay on Friday?

Practice

5C Statistical diagrams

FM **1** The grouped bar chart below shows the lifetimes of some batteries in a test.

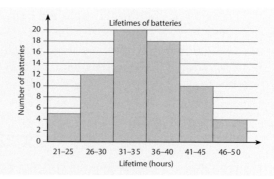

a Make a frequency table for the data.

b What is the modal class?

c What is the total number of batteries tested?

d How many batteries had a lifetime of 40 hours or less?

e What is the maximum and minimum possible range?

FM **2** The dual bar chart shows car sales for two showrooms, Connors and VPW.

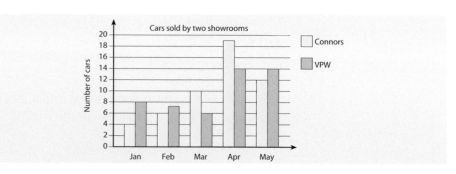

a Which showroom sold the most cars?
b During how many months did Connors sell more cars than VPW?
c What were the maximum numbers of cars sold by each showroom in any month?
d What is the range of cars sold for each showroom?
e What other kind of diagram could illustrate this data?

  3 There are 240 trees in a small wood. The pie chart shows the proportion of each kind of tree.

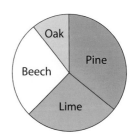

a Of which kind of tree are there fewest?

b Estimate the fraction of pine trees.

c Estimate the number of lime trees.

  4 Jan checks the fuel in her motorbike tank at the end of each day. The line graph below shows the fuel levels for one week.

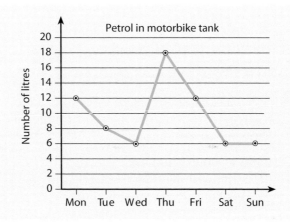

a How much fuel was in the tank at the end of Friday?

b When did the tank contain 6 litres of fuel?

c On which day was the tank refilled?

d What is the range of fuel in the tank for the week?

e On which day did Jan not use her motorbike?

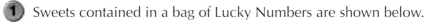
Practice
5D Probability

1 Sweets contained in a bag of Lucky Numbers are shown below.

a One sweet is picked from the bag, at random. What is the probability the number is one of the following?

  i Odd   **ii** Greater than 5   **iii** Odd or even   **iv** A multiple of 3   **v** Not 7

b Which number has a probability of $\frac{1}{5}$ of being picked?

2 Mark Choi has these chopsticks in his kitchen drawer.

Short	Long
6 red	2 red
2 green	8 green
4 yellow	2 yellow

He picks a chopstick at random. Find these probabilities.

a P(red)
b P(green *or* yellow)
c P(short)
d P(long and red)
e P(short and red *or* short and yellow)
f P(not red)

3 Sema has made a Wheel of Fortune game. It costs 5p to play. The numbers on the wheel show the prizes, in pence.

If the wheel is spun once, what is the probability of winning the following prizes? Express your answers as decimals.

a 10p   **b** Nothing
c More than 5p   **d** The biggest prize
e 7p
f Any prize other than 10p

4 a Write down an event with a probability greater than 0 but less than $\frac{1}{2}$.
b Write down an event with a probability greater than $\frac{1}{2}$ but less than 1.

5 An ordinary die is tossed at the same time as this spinner is spun. Their numbers are added together to give the total score.

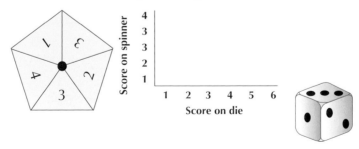

a Draw a sample space diagram to show all the possible outcomes.
b How many possible outcomes are there?
c Find each of the following probabilities.

  i P(score of 8)   **ii** P(score of 6)   **iii** P(even score)   **iv** P(score is not 5)

Practice

5E Experimental probability

1 **a** Toss a die 50 times and record your results in a frequency table.
   b Find the experimental probability of getting 6, writing your answer as:
   i a fraction
   ii a decimal
   c What is the theoretical probability of getting an even number?
   d How many odd numbers would you expect to get after tossing the die 60 times?
   e If you repeated the experiment, would you get the same results?

2 Jason thinks that the probability of getting two heads when tossing three coins is 0.5. Carry out an experiment to see if he is correct.

3 **a** Trace this spinner and cut it out of thin card or paper. Push a drawing pin through the centre.
   b Spin your spinner 50 times. Record your results in a frequency table.
   c Calculate the experimental probability of the spinner landing on a letter.
   d Roughly how many times out of 50 spins would you expect the spinner to land on a number?
   e Make up three events of your own, e.g. 'Lands on even number.' Calculate the experimental probability for each event.

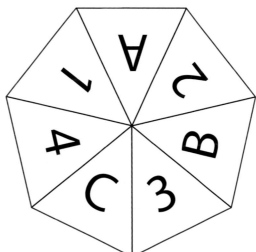

CHAPTER 6 Algebra **2**

Practice

6A Algebraic terms and expressions

1 Write terms or expressions to illustrate the following sentences.
   a Multiply x by four and add three.
   b Multiply y by five and subtract eight.
   c Divide z by three and subtract two.
   d Multiply x by itself and add one.
   e Divide y by two and subtract five.

2 Write down the values of the following terms for each different value of x.

a $5x$ where   **i** $x = 3$ **ii** $x = 10$ **iii** $x = 100$
b x^2 where   **i** $x = 4$ **ii** $x = 7$ **iii** $x = 9$
c $\frac{x}{3}$ where   **i** $x = 6$ **ii** $x = 21$ **iii** $x = 30$
d $0.5x$ where   **i** $x = 10$ **ii** $x = 8$ **iii** $x = 22$

3 Write down the values of these expressions for each different value of t.

a $t^2 - 2$ where   **i** $t = 2$ **ii** $t = 7$ **iii** $t = 10$
b $20 + t^2$ where   **i** $t = 5$ **ii** $t = 0$ **iii** $t = 9$

4 **a** Rearrange these cards to make five different expressions, e.g. $3 + 2 - x$.

b Calculate the value of each of your expressions when $x = 5$.

5 Work out the value of the following expressions for the given values of n.

a $(n + 3)(n - 2)$ where   **i** $n = 4$ **ii** $n = 6$
b $(n - 4)(n + 3)$ where   **i** $n = 6$ **ii** $n = 7$
c $(n - 4)(n - 1)$ where   **i** $n = 5$ **ii** $n = 10$
d $(n + 3)(n + 6)$ where   **i** $n = 4$ **ii** $n = 7$
e $(n - 4)^2$ where   **i** $n = 6$ **ii** $n = 10$

Practice

6B Rules of algebra

1 In each of the following clouds, only two expressions are equal to each other. Write down the equal pair.

$4 + 7$ 7×4 $b - c$ $\frac{c}{b}$

$7 \div 4$ $c + b$

$4 \div 7$ bc $c - b$

4×7 $b + c$ $\frac{b}{c}$

2 In each of the following lists, write down all the expressions that equal each other.

    a 2×5, $5 + 2$, $5 \div 2$, 5×2, $2 - 5$, $2 \div 5$, $2 + 5$, $5 - 2$

    b $\dfrac{t}{3}$, $t - 3$, $3t$, $t \div 3$, $3 + t$, $t \times 3$, $3 - t$

3 Write down two equivalent facts for each of the following facts.

    a $4 \times 6 = 24$ **b** $p - 5 = 7$

    c $10 + 3 = 13$ **c** $\dfrac{x}{2} = 3$

4 Show by substitution of suitable numbers that:

    a   if $a - b = 3$, then $a = b + 3$ **b**   if $pq = 20$, then $\dfrac{20}{p} = q$

5 Show that each of these is true by using suitable substitutions.

    a $2b + c = c + 2b$ **b** $3ab = 3ba$

6 a^2 means $a \times a$ or simply aa.
So, a^2b means $a \times a \times b$ or simply aab.
Write down three expressions equivalent to each of the following.

    a a^2b **b** x^2y^2 **c** $10m^2$

Practice

6C   Simplifying expressions

1 Simplify each of the following expressions.

    a $3p + 6p$ **b** $12q - 4q$ **c** $5t + 3t + 8t$
    d $4a + 3a - 7a$ **e** $9x - x$ **f** $20c - 6c - 6c$
    g $8m - m - 6m$ **h** $20k - 13k + 2k$ **i** $y + 17y - 9y$

2 Simplify each of the following expressions.

    a $2a + 3a + 5b + 6b$ **b** $5x + 3y + 3x + y$
    c $7d - 3d + 8e - 5e$ **d** $2 + 6m - 3m + 7$
    e $4p + 7q - 2p + 3q$ **f** $10t + 15 - 10 - 6t$
    g $5f + 7p - 2f + 3$ **h** $12k + 9m - 2m - 12k$
    i $9t + 9 - 3t - 7$

3 Expand the brackets.

    a $5(d + 3)$ **b** $2(5s + 3)$ **c** $10(a - 1)$
    d $4(6m - 3)$ **e** $20(a + b - c)$ **f** $4(3k - 2 + 4x)$
    g $6(10 - 2p - 3r)$

4 Expand and simplify each of the following expressions.

    a $2(x + 3) + 3(x + 5)$ **b** $4(m + 1) + 7(m + 2)$
    c $5(a + 2) + 2(a - 3)$ **d** $6(3d + 4) + 2(7d + 2)$
    e $2(3p + 8) + 3(2p - 1)$ **f** $8(3t + 4) + 7(2t - 3)$

5 Expand and simplify each of the following expressions.

a $2(x + 1) + 3(x - 2)$　　　　　**b** $4(x - 2) - 3(x - 1)$
c $5(x + 3) - 3(x + 4)$　　　　　**d** $6(x - 3) + 2(x - 5)$
e $2(3x - 1) - 3(x - 3)$　　　　　**f** $4(2x - 1) - 3(x - 5)$

Practice

6D Formulae

FM **1** Write each of these rules as a formula.

a The total distance, t, covered by running a number of laps, n, of a 400 m track.

b The cost of one loaf of bread, b, if 5 loaves cost m pence.

c The difference in height between Sven, 180 cm tall, and Tammy, who is shorter, is d. Tammy is T cm tall.

d The total weight of a box weighing 100 grams and an electric motor, is w. The motor weighs m grams.

FM **2** The formula shows the time taken (minutes) for a machinist to make a number of wooden chair legs

$$t = 15 + 7n$$

where t = total time taken, n = number of chair legs. Find the time taken when:

a $n = 3$　　　　　**b** $n = 50$　　　　　**c** $n = 17$

FM **3** The number of workers it takes to complete a job is given by the formula below

$$N = \frac{180}{t}$$

where N = number of workers, t = time spent by each worker on the job. Find the number of people working on the job when:

a $t = 60$ hours　　　　**b** $t = 4$ hours　　　　**c** $t = 15$ hours

FM **4** The formula below gives the total cost of 15 plugs and 15 sockets

$$C = 15(P + S)$$

where C = total cost, P = price of a plug (in pence), S = cost of a socket (in pence). Calculate the total cost when:

a $P = 40$p, $S = 260$p　　　　　**b** $P = 25$p, $S = 100$p
c $P = 32$p, $S = 172$p

FM **5** The formula below gives the area of cardboard (cm²) needed to make a box of length 15 cm

$$A = 15w - 4x$$

where A = area, w = width of box, x = side of square cut-out. Calculate the area of cardboard needed when:

a $w = 10\,\text{cm}$, $x = 3\,\text{cm}$ **b** $w = 20\,\text{cm}$, $x = 5\,\text{cm}$

c $w = 17\,\text{cm}$, $x = 4\,\text{cm}$

Practice

6E Equations

1 Solve each of the following equations.

a $5x = 20$ **b** $10y = 10$ **c** $8p = 32$ **d** $9d = 72$
e $2q = 300$ **f** $7n = 154$ **g** $3r = 81$ **h** $4t = 240$

2 Solve each of the following equations.

a $p + 5 = 12$ **b** $m + 8 = 17$ **c** $b + 30 = 100$
d $h + 16 = 32$ **e** $d - 3 = 8$ **f** $p - 9 = 12$
g $m - 23 = 17$ **h** $s - 40 = 170$

3 Solve each of the following equations.

a $2y + 1 = 11$ **b** $3b + 2 = 14$ **c** $7x + 6 = 41$
d $10p + 40 = 100$ **e** $6q + 8 = 56$ **f** $2m + 100 = 400$
g $5d + 17 = 52$ **h** $8r + 28 = 92$

4 Solve each of the following equations.

a $4m - 3 = 13$ **b** $2q - 7 = 11$ **c** $5v - 10 = 25$
d $7n - 13 = 8$ **e** $10a - 60 = 100$ **f** $2w - 35 = 35$
g $6f - 18 = 36$ **h** $11x - 17 = 49$

5 Two of these equations have the same solution. Find them. Write down the solutions to all of the equations.

a $5x + 2 = 22$

b $3x - 5 = 22$

c $9x + 20 = 65$

d $4x - 11 = 33$

e $10x + 15 = 105$

f $x + 97 = 123$

g $6x + 7 = 25$

h $5x - 20 = 50$

Solve each of the following equations. The answers may be decimals or fractions.

a $2x - 4 = 9$ **b** $4x + 6 = 13$ **c** $5x - 2 = 9$ **d** $10x + 4 = 18$

CHAPTER 7 Geometry and Measures **2**

Practice   7A Lines and angles

1 Describe each of the following angles as acute, right-angle, obtuse or reflex. Estimate the size of each angle.

a   **b**   **c**

d   **e**   **f**

2 Write down four different properties of shape ABCDEF.

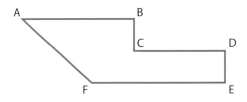

3 Write down the geometric properties of this kite.

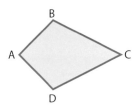

4 For each part, mark any equal sides, parallel sides, right-angles, etc.

   a Sketch a quadrilateral with exactly two parallel lines and two pairs of perpendicular sides.

   b Sketch a pentagon (five-sided shape) with exactly three equal sides and one right angle.

   c Sketch a hexagon (six-sided shape) with exactly two parallel lines, three obtuse angles and two acute angles.

5 **a** Write down three pairs of corresponding angles.

   b Write down three pairs of alternate angles.

   c Which of these statements are true?

     i *l* and *d* are equal.

     ii *c* and *j* are corresponding angles.

     iii *p* and *f* are alternate angles.

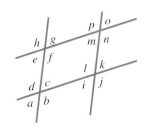

Practice

7B Calculating angles

Calculate the size of each unknown angle.

1 **a** 280° 32° *a*

   b 145° *b*

   c 85° *c*

   d 43° *d*

2 **a** 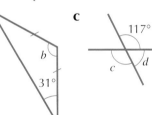 53° *a*

   b *b* 31°

   c 117° *c* *d*

   d 20° 40° 30° *e* 130°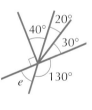

3 **a** 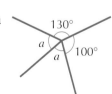 130° *a* 100° *a*

   b 140° *b* 78°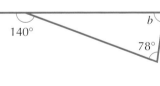

4 **a** 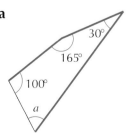 30° 165° 100° *a*

   b *b* 42°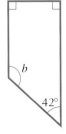

   c 114° *c* 123° 132° *d*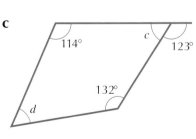

4

1 Write down the coordinates of the points A, B, C, D and E.

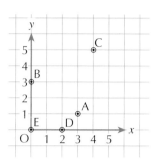

2 a Make a copy of the grid in question 1 and plot the points A(1, 1), B(3, 1) and C(5, 4).
   b The three points are the vertices of a parallelogram. Plot the point D to complete the parallelogram.

3 Write down the coordinates of the points A, B, C, D, E, F and G.

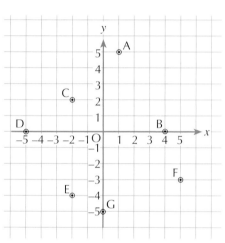

4 a Make a copy of the grid in question 3 and plot the points A(−5, −3), B(−5, 4), C(2, 4).
   b Join the points to make a triangle.
   c Plot point D so that ABCD is a square. Draw the other two sides of the square.
   d Each diagonal of the square crosses the axes. Write down the coordinates of these intersections.

CHAPTER **8** Statistics **2**

5

Local residents were asked how they would like to see some waste ground developed. The results are shown in the tally chart below.

Development	Tally	Frequency
Swimming pool	𝍱 /	
Playground	𝍱 𝍱 //	

Supermarket	⅂⊬⊬⊬ ////
Health centre	⅂⊬⊬⊬
Cafe	///

 1 Draw a bar chart to illustrate the data.

 2 Write possible reasons why the residents voted for each kind of development.

Practice
8B Using the correct data

You are going to investigate the number of words per line in a paperback novel. Follow these steps:

1 Copy this tally chart. You will need 10 to 15 rows.

Number of words	Tally	Frequency
1		
2		
3		
4		

2 Find a paperback novel. Choose a page at random. Count the number of words on each line. Do this for 50 lines. Record the data in your tally chart.

3 Decide how to count hyphenated words, numbers, etc.

4 Draw a bar chart to illustrate your data.

5 If you have time, repeat the activity using a different paperback. Compare your results.

Practice
8C Grouped frequencies

1 Some pupils were asked how many music CDs they owned. The results are shown below.

14	1	4	7	0	11	7	2	17	9
20	6	10	0	5	3	21	0	15	6
11	0	13	3	8	2	18	9	27	1
12	8	1	18						

a   Copy and complete this grouped frequency table.

Number of CDs	Tally	Frequency
0–4		
5–9		

b   Draw a bar chart to illustrate the data.

2   The average ages of families at a small holiday resort are shown below.

31	17	31	21	32	29	27	39	25	25
40	27	36	28	46	19	38	32	23	28
35	19	41	30	24	34	55	26	20	36
43	51								

a   Which is the most sensible table for the data? Explain your answer.

Average age	Tally	Frequency
10 ≤ Age < 15		
15 ≤ Age < 20		

Average age	Tally	Frequency
0 ≤ Age < 20		
20 ≤ Age < 40		

Average age	Tally	Frequency
10 ≤ Age < 20		
20 ≤ Age < 30		

b   Copy and complete the table you chose in part **a**.
c   Draw a bar chart to illustrate the data.

8D Data collection

Kickoff Ltd closes down during the summer holiday. Office and factory staff were asked these questions about their preferred holiday times.

'How many days of your annual leave would you like to use for your summer holiday?'
'In which month would you like your summer holiday?'
'On which day of the week should the holiday begin?'

Their answers are shown in the table below.

Employee	Male/Female	Days of annual leave	Month	Weekday
Office	M	10	June	Monday
Office	F	5	August	Monday
Factory	M	5	August	Wednesday
Office	M	10	June	Monday
Factory	F	8	July	Wednesday
Factory	M	12	August	Thursday
Factory	M	10	July	Monday
Office	F	10	August	Monday
Factory	F	5	July	Wednesday
Factory	F	15	July	Monday
Factory	M	10	June	Monday
Factory	F	5	August	Monday
Factory	M	10	August	Monday
Office	F	12	July	Thursday
Factory	F	10	August	Monday
Factory	M	5	August	Monday
Factory	F	5	July	Friday
Office	F	10	June	Monday
Office	M	5	August	Wednesday
Factory	F	10	July	Monday

1 **a** Make a frequency table for the number of days chosen by office staff.
   b Make a frequency table for the number of days chosen by factory staff.
   c Write a sentence comparing office and factory staff.

2 **a** Make a frequency table for the holiday months chosen by male staff.
   b Make a frequency table for the holiday months chosen by female staff.
   c Write a sentence comparing male and female staff.

3 **a** Make a frequency table for the start day chosen by office staff.
   b Make a frequency table for the start day chosen by factory staff.
   c Write a sentence comparing office and factory staff.

4 Think of another question about staff holidays. Write it down as you would on a data collection sheet, using answer boxes.

CHAPTER 9 Number 3

9A Rounding

1 Round off these numbers to the following.
   i The nearest 10   **ii** The nearest 100   **iii** The nearest 1000

   a 657   **b** 2555   **c** 3945   **d** 409   **e** 17 059 **f** 9895

2 **a** What are these Test Your Strength scores, to the nearest 100?
   b Estimate the scores to the nearest 10.

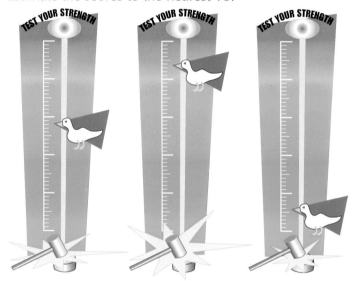

3 **a** Read these weighing scales to the nearest gram.
   b Read the scales to the nearest 10 g.
   c Read the scales correct to one decimal place.

grams

7.6   7.7   7.8   7.9

24.2   24.3   24.4   24.5

grams

0.8   0.9   .0   1.1

grams

4 Round off these numbers to the following.
   i The nearest whole number   **ii** One decimal place

   a 7.32   **b** 8.75   **c** 3.04   **d** 19.58   **e** 0.749   **f** 9.955

5 Round these bank balances using a sensible degree of accuracy. Then place them in order of size, starting with the smallest.

Andrew	£4960.78	Jamil	£680	Geoff	£4799.50
Jean	£12 600	Suzy	£6709.67	Laurence	£94.45
Harry	£6921.44	Sema	£4934.24	Darren	£5094.68

9B The four operations

Show your working for each question.

FM **1** Calculate the total cost of 35 burger meals at £3.98 each.

FM **2** Electric light bulbs can be packed into boxes of 16 or 24. How can 232 bulbs be packed into full boxes only? Find two possible ways.

FM **3** Which is better value?

Discount Plus — 13 videos only £25

Bargain Basement — 11 videos only £20

4 **a** Find two consecutive even numbers whose product is 288.

b Find three consecutive odd numbers whose sum is 111.

FM **5** A postmaster has a full sheet of 80 blue stamps costing 27p each. He also has a full sheet of 90 red stamps costing 19p each.

a What is the total value of the two sheets of stamps?

b He sells 31 blue stamps. What is the value of the rest of that sheet?

c How many red stamps can be bought for £5?

d How many blue stamps can be bought for £10?

6 A full bottle contains between 80 ml and 90 ml of medicine. If it were used for 5 ml doses, there would be 3 ml left in the bottle. If it were used for 3 ml doses, there would be 1 ml left in the bottle. How much medicine is in a full bottle?

9C BODMAS

1 Circle the operation that you do first in these calculations, then work it out.

a $9 - 6 \div 3$ **b** $3 \times (20 - 15)$ **c** $50 \div 10 - 5$

d $(3 + 13) \div 4$

2 Work out each of the following, showing each step of the calculation.

a $20 + 20 \div 4$ **b** $4 \times 8 - 3 \times 7$ **c** $(8 - 2) \times 4$

d $6 + 4^2$

3 Put brackets into each of these statements to make the calculation true.

a $20 - 10 - 3 = 13$ **b** $5 + 2 \times 2 = 14$
c $9 - 7 - 5 - 3 = 4$ **d** $5 - 2^2 = 9$
e $12 - 3 \times 10 - 6 = 0$ **f** $10 + 10 \div 10 + 10 = 1$
g $1 + 3^2 + 5^2 = 41$ **h** $40 - 20 \div 4 + 1 = 36$

4 Work out the value of each of the following.

a $5^2 - 5$ **b** $(5 + 5) \div 5 + 5$ **c** $5 - 5 \times (5 - 5)$
d $5 \times 5 - (5 \div 5) \times (5 + 5)$

5 Write down each of the following using a single calculation. Use numbers and symbols +, −, ×, ÷, ( ) only. Then calculate the answer.

a Subtract the product of 4 and 2 from 12.
b Multiply the sum of 9 and 8 by 3.
c Add the square of 3 to 5.
d Subtract the square of 3 from the square of 4. Then subtract the answer from 20.

Practice

9D Long multiplication and long division

1 Work out each of the following long multiplication problems. Use any method you are happy with.

a 13×32 **b** 54×27 **c** 19×275 **d** 148×38

2 Work out each of the following long division problems. Use any method you are happy with. Some of the problems will have a remainder.

a $420 \div 12$ **b** $600 \div 22$ **c** $738 \div 38$ **d** $884 \div 26$

Decide whether these problems are long multiplication or long division. Then do the appropriate calculation, showing your method clearly.

3 A sports stadium has 23 rows of seats. Each row has 84 seats. How many people can be seated in the stadium?

4 A bag contains 448 g of flour. The flour is used to make cakes. Each cake contains 14 g of flour.

a How many cakes were made?
b How much flour is needed to make 234 cakes?

5 Floor tiles measure 35 cm by 26 cm. They cover a floor measuring 945 cm by 650 cm.

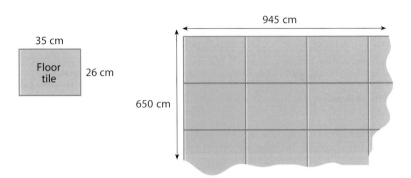

a How many tiles are next to the long edge of the floor?

b How many tiles are next to the short edge of the floor?

c How many tiles were used to cover the floor?

Practice

9E Efficient calculations

1 Without using a calculator, work out the value of each of the following.

   a $\dfrac{20 - 6}{2 + 5}$ **b** $\dfrac{7 + 9}{0.9 - 0.4}$

2 Use a calculator to do the calculations in question 1. Did you get the same answers as before?

3 For each part of question 1, write down the sequence of keys that you pressed.

4 Work out the value of each of the following. Round off your answers to one decimal place, if appropriate.

   a $\dfrac{689 + 655}{100 - 58}$ **b** $\dfrac{420 - 78}{54 \div 3}$ **c** $\dfrac{36 \times 84}{29 + 17}$ **d** $\dfrac{296 + 112}{183 - 159}$

5 **a** Estimate the answer: $\dfrac{443 + 178}{53 - 27}$

   b Now use a calculator to work out the answer to one decimal place. Is your answer about the same?

6 Calculate each of the following.

   a $\sqrt{344\,569}$ **b** 5.4^2

   c $\sqrt{(50.6 + 39.65)}$ **d** $(12.3 - 2.6)^2$

7 **a** Calculate: $\dfrac{387 + 231}{306 \div 17}$

   b Write your answer to part **a** as a mixed number.

9F Calculating with measurements

1 Convert each of the following measurements.

a	160 mm to cm	**b**	0.53 m to cm	**c**	0.035 km to m
d	9.7 m to cm	**e**	5200 m to km	**f**	37 m to km
g	4000 cm to km	**h**	23 cm to mm	**i**	8.25 m to mm
j	9 km to cm	**k**	3400 g to kg	**l**	9.32 kg to g
m	265 g to kg	**n**	0.01 kg to g	**o**	9 g to kg
p	3200 ml to l	**q**	5.729 l to ml		

r 115 minutes to hours and minutes

s 924 minutes to hours and minutes

2 Add together each set of measurements and give the answer in an appropriate unit.

a 0.063 kg, 580 g, 0.4 kg **b** 450 ml, 0.63 l, 9 cl

c 640 mm, 94 cm, 0.003 km

 3 Fill in each missing unit.

a A car weighs 1262 …

b A £1 coin is about 3 … thick.

c A wine glass holds about 30 … of wine.

d Scott ran 200 … in a time of 32 …

CHAPTER 10 Algebra 3

10A Square numbers and square roots

1 Write down the value of each of the following. (Do not use a calculator.)

a $\sqrt{64}$ **b** $\sqrt{1}$ **c** $\sqrt{121}$ **d** $\sqrt{81}$ **e** $\sqrt{0}$

2 With the aid of a calculator, write down the value of each of the following.

a $\sqrt{196}$ **b** $\sqrt{784}$ **c** $\sqrt{8281}$ **d** $\sqrt{344\,569}$

e $\sqrt{3\,364\,000\,000}$

3 Make an estimate of each of the following square roots, then use a calculator to see how many you got right.

a $\sqrt{169}$ **b** $\sqrt{484}$ **c** $\sqrt{900}$ **d** $\sqrt{3600}$ **e** $\sqrt{2209}$

4 Sometimes, the differences between two square numbers is another square number. For example:

$$10^2 - 8^2 = 100 - 64 = 36 \text{ and } 6^2 = 36, \text{ so } 10^2 - 8^2 = 6^2$$

Use the numbers in the cloud to find more of these. Write each answer like this: $10^2 - 8^2 = 6^2$.

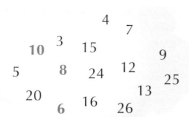

5 **a** Copy and continue the pattern to make eight rows. Work out all of the answers.

$$
\begin{aligned}
1 &= \\
1 + 3 &= \\
1 + 3 + 5 &= \\
1 + 3 + 5 + \ldots &=
\end{aligned}
$$

b What can you say about the answers?
c Can you find a rule that gives the answer? Check that your rule works.

6 Write down the full solution to each of these equations.

a $x^2 = 64$ **b** $x^2 = 81$ **c** $x^2 + 5 = 30$ **d** $x^2 - 10 = 111$

7 Write down the full solution to each of these equations.

a $4x^2 = 16$ **b** $6x^2 = 6$ **c** $3x^2 - 35 = 73$

Practice

10B Triangle numbers

1 Look at the numbers in the box. Write down the numbers from the box that are the following.

a Square numbers   **b** Triangle numbers
c Odd numbers   **d** Multiples of 3
e Factors of 48   **f** Prime numbers

1, 2, 3, 4, 5, 6, 13, 16, 18, 21, 22, 36, 40, 41, 45

2 Each of the following numbers is the sum of two triangle numbers. Write them down, e.g. 16 = 1 + 15.

a 4   **b** 18   **c** 38   **d** 42   **e** 70

3 Every odd number can be written as the difference between two triangle numbers, e.g. 9 = 15 − 6.

Express the following odd numbers as the difference between two triangle numbers.

a 5   **b** 11   **c** 21   **d** 29

④ Write down the first 20 triangle numbers. Cross out the numbers in these positions: 2nd and 3rd, 5th and 6th, 8th and 9th and so on. Calculate the differences for the remaining sequence. Describe the differences.

10C From mappings to graphs

① For each of the following functions, do the following.
i Complete the arrow diagram.
ii Complete the coordinates alongside.
iii Plot the coordinates and draw the graph of the function.

a $y = 4x$ **b** $y = 3x + 4$

x	$y = 4x$	Coordinates
0 →	0	(0, 0)
1 →	4	(1, 4)
2 →		(2,)
3 →		(3,)
4 →		(4,)
5 →		(5,)

x	$y = 3x + 4$	Coordinates
0 →	4	(0, 4)
1 →	7	(1, 7)
2 →		(2,)
3 →		(3,)
4 →		(4,)
5 →		(5,)

c $y = 3x - 1$

x	$y = 3x - 1$	Coordinates
1 →	2	(1, 2)
2 →		(2,)
3 →		(3,)
4 →		(4,)
5 →		(5,)
6 →		(6,)

② For each of the following functions, do the following.
i Choose some of your own starting points.
ii Make your own mapping diagram.
iii Draw a graph.

a $y = x + 5$ **b** $y = 2x + 7$ **c** $y = 5x - 3$ (Hint: Start with $x = 1$.)

10D More about graphs

① **a** Draw the following graphs on the *same* grid, and label them.
Use these axes: x-axis from 0 to 7, y-axis from 0 to 7.
i $y = 3$ **ii** $y = 5.5$ **iii** $x = 6$ **iv** $x = 0.5$

b Write the coordinates of the point where the following lines cross each other. Do not draw their graphs.

   i $y = 4$ and $x = 1$ **ii** $y = 2.5$ and $x = 10$

2 **a** For each of the functions **i** to **iv**, complete the arrow diagram, make a set of coordinates and plot the points on the *same* grid.

    $x \rightarrow$? **Coordinates**

    $-2 \rightarrow$ $(-2, \quad)$
    $-1 \rightarrow$ $(-1, \quad)$
     $0 \rightarrow$ $(0, \quad)$
     $1 \rightarrow$ $(1, \quad)$
     $2 \rightarrow$ $(2, \quad)$
     $3 \rightarrow$ $(3, \quad)$
     $4 \rightarrow$ $(4, \quad)$
     $5 \rightarrow$ $(5, \quad)$

    Use these axes: x-axis from -2 to 5, y-axis from -6 to 6.

    i $y = x$ **ii** $y = x - 1$ **iii** $y = x - 2$ **iv** $y = x - 3$

b Explain what you notice.

c Draw the graph of $y = x - 4$ without calculating any coordinates.

3 Draw the graph of the function $y = 2x + 5$. Use these axes: x-axis from -2 to 5, y-axis from 0 to 15.

Practice

10E Questions about graphs

Try answering the first three questions without drawing graphs.

1 **a** Is the point $(3, 6)$ on the graph of $y = 2x$?
   b Is the point $(20, 16)$ on the graph of $y = x - 6$?
   c Is the point $(5, 17)$ on the graph of $y = 3x + 2$?
   d Is the point $(9, 76)$ on the graph of $y = 10x - 14$?

2 Which of the following lines does the point $(5, 9)$ lie on?

   $y = 4x$ $y = x + 3$ $y = 2x - 1$ $y = 5$

3 Write down two functions whose graphs will pass through the point $(1, 5)$.

4 **a** Draw the graphs of $y = 2x + 3$ and $y = 12 - x$ on the *same* axes. Use these axes: x-axis from 0 to 5, y-axis from 0 to 15.
   b Write down the coordinates of the point of intersection.

5 Which of the points $(2, 9)$, $(4, 18)$, $(3, 13)$ is the intersection of the graphs $y = 4x + 1$ and $y = 5x - 2$? **Do not draw the graphs.**

CHAPTER **11** Geometry and Measures **3**

11A Measuring and drawing angles

1 Measure the size of each of the following angles, giving your answers to the nearest degree.

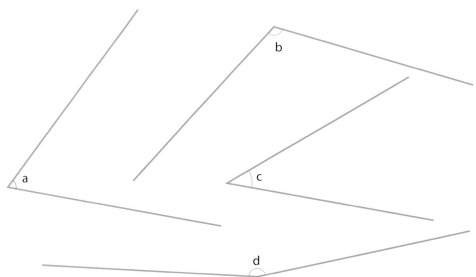

2 Draw and label each of the following angles.

    a 70°   **b** 42°   **c** 134°   **d** 200°   **e** 343°

3 **a** Measure all the angles in the quadrilateral ABCD.
   b Add the angles together.
   c Comment on your answer.

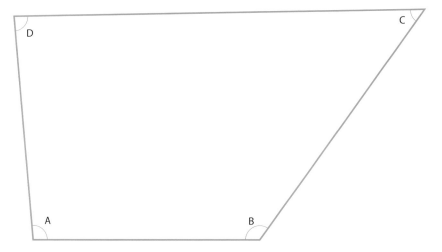

11B Constructions

1 Construct the following triangles. Remember to label the vertices and angles.

a

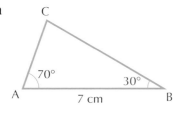

b

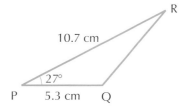

   c On your drawing, measure sides BC and AC, to the nearest millimetre.
   d On your drawing, measure ∠Q and ∠R, to the nearest degree.

2 **a** Construct the triangle ABC with AB = 6.4 cm, BC = 11.3 cm, ∠B = 16°.
   b Measure side AC, to the nearest millimetre.
   c Measure angles ∠A and ∠C, to the nearest degree.

3 **a** Construct the quadrilateral PQRS.
   b On your drawing, measure ∠R and ∠S, to the nearest degree.
   c On your drawing, measure RS, to the nearest millimetre.

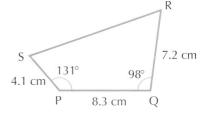

4 Draw the line AB 9 cm long. Construct the perpendicular bisector of AB.

5 Draw an angle of 150°. Construct the bisector of the angle.

11C Solving geometrical problems

1 Draw the different types of quadrilateral. Draw one diagonal for each quadrilateral. Describe the triangles you have made.

2 Explain the difference between a parallelogram and a trapezium.

3 How many different kinds of quadrilateral can be constructed on this pin-board?

Use square dotted paper to record your quadrilaterals.
Beneath each quadrilateral, write its name.

4 A square is a special kind of rectangle. Fill in the blanks to make different statements.

   a A parallelogram is a special kind of …
   b A … is a special kind of parallelogram.
   c A square is a special kind of …
   d A … is a special kind of kite.

CHAPTER **12** Number **4**

5

6

Practice

12A Percentages

1. Without using a calculator, work out each of these.

    a 20% of 70   **b** 21% of 40   **c** 32% of 24   **d** 76% of 400

2. Work out each of these using a calculator.

    a 95% of 220 kg   **b** 31% of 6 m   **c** 53% of 72 years
    d 11% of 17 litres   **e** 1% of 723 pages   **f** 99% of 634 houses
    g 6% of 4200 seeds   **h** 73% of £2427

3. Which is bigger?

    a 19% of £73 *or* 32% of £56   **b** 9% of 523 g *or* 99% of 46 g

4. Write down or work out the equivalent percentage and decimal to each of these fractions.

    a $\frac{3}{4}$ **b** $\frac{4}{5}$ **c** $\frac{17}{20}$ **d** $\frac{9}{25}$ **e** $\frac{7}{8}$

5. Write down or work out the equivalent percentage and fraction to each of these decimals.

    a 0.4   **b** 0.9   **c** 0.27   **d** 0.65   **e** 0.62

6. Write down or work out the equivalent fraction and decimal to each of these percentages.

    a 35%   **b** 60%   **c** 28%   **d** 17.5%   **e** 43%

7. Arrange these numbers in increasing order of size.
    (Hint: Convert the numbers to percentages.)

    a 0.65, $\frac{16}{25}$, 63%   **b** 0.09, $\frac{1}{8}$, 11%

Practice

12B Ratio and proportion

1. Five reels of fishing line have a total length of 72 m. Find the total length of the following.

    a 3 reels   **b** 11 reels

2. The table shows the seeds contained in four packets of Summer Border.

4

Flower	Number of seeds
Cloth of Gold	20
Delphinium	12
Penstemon	8
Mrs Perry	6

Calculate the number of each seed contained in the following.

a 12 packets   **b** 6 packets

3 For each of these shapes, work out the following.
a What proportion is shaded?
b What is the ratio of the shaded part to the unshaded part?

i 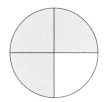   **ii** 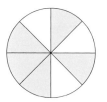   **iii**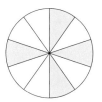

4 The contents of two boxes of muesli are shown below.

a What is the total weight of each box?

b Copy and complete the table below, which shows the proportion of each ingredient.

	Oats	Wheat	Nuts	Fruit
Luxury Muesli	25%			
Natural Breakfast				10%

5 In an orchestra, the ratio of cellists to violinists is 3 : 5. There are 20 violinists in the orchestra. How many cellists are there?

Practice

12C Calculating ratios and proportions

1 Reduce each of the following ratios to its simplest form.

a 9 : 6   **b** 8 : 20   **c** 24 : 30   **d** 200 : 900
e 75 : 45   **f** 21 : 56

2 Write down the ratio of coloured to white squares for each of these grids.

a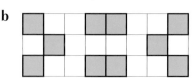

b

3 Four football teams scored 250 goals altogether. Beetwood Rangers scored 50 goals, Trenton FC scored 35, Middleton Wanderers scored 90 and Pinkhams scored the rest.

   a Write down the percentage of the total goals that each team scored.
   b Write down each of the following goal ratios in its simplest form.
      i Beetwood Rangers : Trenton FC
      ii Middleton Wanderers : Pinkhams
      iii Pinkhams : Beetwood Rangers

4 The atmosphere of Uranus is made up of hydrogen and other gases in the ratio 6 : 1. A space probe filled two flasks with samples of the atmosphere.

   a One flask contained 18 litres of hydrogen. How much other gases did it contain?
   b One flask contained 0.8 litres of other gases. How much hydrogen did it contain?

Practice

12D Solving problems

1 Divide £180 in each of these ratios.

   a 4 : 5   **b** 11 : 4   **c** 5 : 7   **d** 22 : 23

2 There are 96 houses on an estate. The ratio of detached to semi-detached houses is 11 : 5. How many of each kind of house are there?

3 Andrea divides up her free time in the ratio:

   sewing : reading : photography = 7 : 6 : 12

She has 300 minutes of free time. How much time does she spend on each activity?

4 Dominic has 30 coloured sweets left in a packet. These are in the ratio:

   red : green = 2 : 3

   a How many of each colour are there?
   b If he eats 2 red sweets and 3 green sweets, what is the new ratio red : green? Comment on your answer.

5 Share 400 kg in each of these ratios.

   a 1 : 2 : 5   **b** 6 : 11 : 3

Algebra **4**

13A Solving 'brick wall' problems

Find the unknown number x in each of these 'brick wall' problems.

1
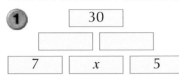

2
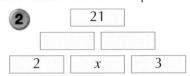

3
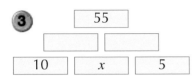

4
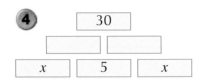

5
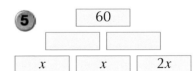

6
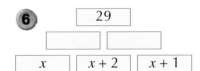

13B Square-and-circle problems

Find the solution set to each of the following square-and-circle puzzles. All solutions must use positive numbers.

1
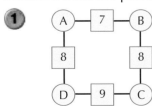

2
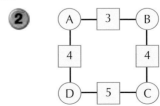

3
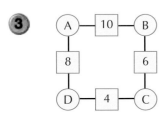

4
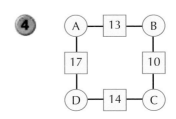

1 Use algebra to solve the triangle-and-circle problems. All the solutions have positive integers.

a
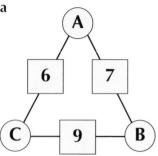

b
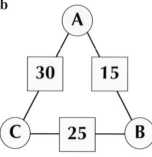

c
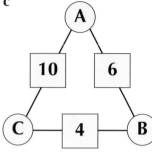

2 Use algebra to solve the triangle-and-circle problems. The solutions involve positive and negative integers.

a
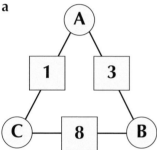

b
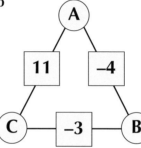

c
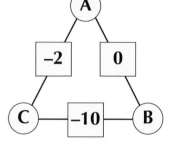

CHAPTER **14** Geometry and Measures **4**

1 Copy each of these shapes and draw its lines of symmetry. Write below each shape the number of lines of symmetry it has.

a

b

c

d

e

2 Write down the number of lines of symmetry for each of the following shapes.

a   b   c   d

3 **a** Copy each shape onto squared paper. Draw an extra line to make the shape have one line of symmetry.

i   ii   iii

b Copy each shape onto squared paper. Draw extra lines to make the shape have two lines of symmetry.

i   ii   iii

4 Copy each of these shapes. Write the order of rotational symmetry beneath the shape. Mark the centre of rotation.

a   b   c   d

5 Write down the order of rotational symmetry for each of the following shapes.

a   b   c   d

6 **a** Copy each shape onto squared paper. Shade an extra square to give the shape rotational symmetry, order 2. Mark the centre of rotation.

i   ii   iii

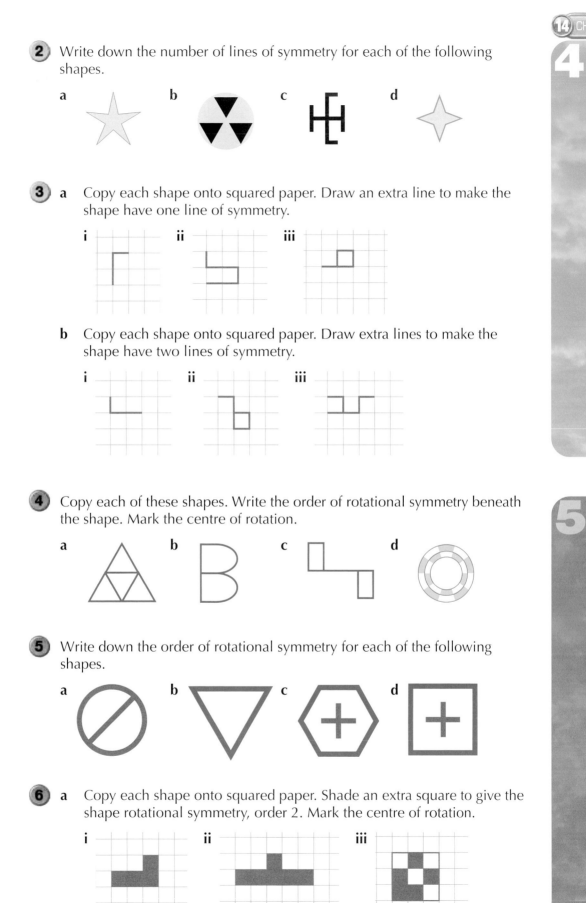

b Copy each shape onto squared paper. Shade extra squares to give the shape rotational symmetry, order 4. Mark the centre of rotation.

i 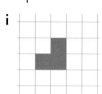 **ii** 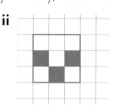 **iii**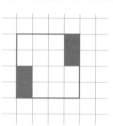

7 Draw new shapes on squared paper with the following properties.

a Rotational symmetry, order 4.
b Rotational symmetry, order 2, with no line symmetry.
c Rotational symmetry, order 2, with two lines of symmetry.
d Rotational symmetry, order 4, with four lines of symmetry.

Practice

14B Reflections

1 Copy each of these diagrams onto squared paper and draw its reflection in the given mirror line.

a 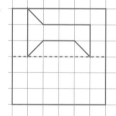 **b**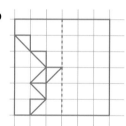

c 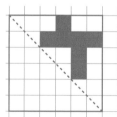 **d**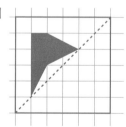

2 The diagrams show the reflections of numbers (images). Copy each diagram. Draw the numbers (objects).

a **b** 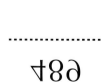 **c**

3 Look at the points shown on the grid below.

a Copy the grid onto squared paper and plot the points A, B, C and D. Include the coordinates on your diagram. Draw the mirror line.

b Reflect the points in the mirror line and label them A', B', C' and D'.

c Write down the coordinates of the image points.

d The point E' (4, 5) is the reflection of point E. Mark both points E and E' on your diagram, including the coordinates.

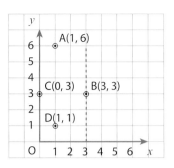

Practice

14C Rotations

1 Copy each of the shapes below onto a square grid. Draw the image after each one has been rotated about the point marked X through the angle indicated. Use tracing paper to help.

a

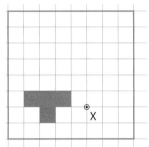

90° clockwise

b

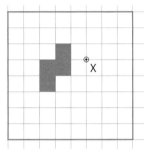

180°

c

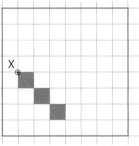

90° anticlockwise

d

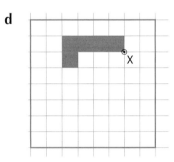

90° anticlockwise

2 **a** Rotate the triangle ABC through 90° anticlockwise about the point (4, 4) to give the image A'B'C'.

b Write down the coordinates of A', B' and C'.

c Which coordinate point remains fixed throughout the rotation?

d Fully describe the rotation that will map the triangle A'B'C' onto the triangle ABC.

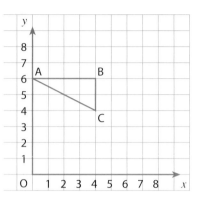

3

| 90° clockwise | 180° | 90° anticlockwise |

a Copy each number.

b Rotate the number through the given angle. Use the dot as the centre of rotation.

14D Translations

 1 Describe each of the following translations.

a A to C
b A to D
c C to B
d D to E
e B to A
f B to D
g D to C

2 Copy the grid and kite ABCD onto squared paper. Label the kite M.

a Write down the coordinates of the vertices of kite M.

b Translate kite M 2 units left, 6 units down. Label the new kite P.

c Write down the coordinates of the vertices of kite P.

d Translate kite P 5 units left, 6 units up. Label the new kite Q.

e Write down the coordinates of the vertices of kite Q.

f Translate kite Q 1 unit right, 3 units down. Label the new kite R.

g Write down the coordinates of the vertices of kite R.

h Describe the translation that maps kite R onto kite S.

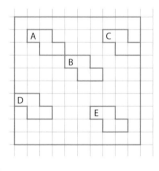

1 Copy each shape and enlarge it by the given scale factor. Use the centre of enlargement O.

a   O ×   **b**

Scale factor 2   Scale factor 3

2 Copy the diagram onto squared paper and enlarge it by a scale factor of 2. Use the point (0, 2) as the centre of enlargement.

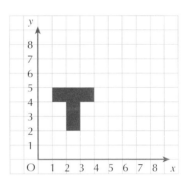

3 On squared paper, draw axes for x and y from 0 to 10. Plot the points A(2, 1), B(3, 3), C(0, 2). Join the points to form the triangle ABC. Enlarge the triangle by a scale factor of 3, using the origin as centre of enlargement. Label your enlargement A'B'C'.

CHAPTER 15 Statistics 3

1 Draw pie charts to represent the following data.

a Favourite computer game of 40 players.

Computer game	No. of players
Wings	6
Venture II	12
Armageddon	8
Space Dust	4
Truckers	10

b   Number of bedrooms in 60 family homes.

Bedrooms	1	2	3	4	5
Homes	7	16	24	9	4

2   The pie chart shows the kinds of tropical fish in an aquarium. There are 24 fish in the aquarium.

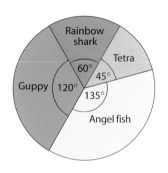

a   How many guppies are there?
b   How many rainbow sharks are there?
c   How many tetras are there?
d   How many angel fish are there?

Practice

15B   Comparing data

1   The data below shows pizza delivery times, in minutes, for two rival companies.

Pizza-to-Go	16, 22, 18, 19, 22, 17, 17, 19, 19, 21, 18, 20
Whizza Pizza	14, 20, 13, 16, 15, 27, 19, 22, 12, 15, 17, 14

a   Calculate the mean delivery time for Pizza-to-Go.
b   Calculate the mean delivery time for Whizza Pizza.
c   Calculate the range for Pizza-to-Go.
d   Calculate the range for Whizza Pizza.
c   Which company would you use? Explain your answer.

2   Class 7A has to decide which person will represent them in the annual School Brain competition. The results of the top two children in a range of tests are shown below.

James	7, 9, 6, 7, 5, 1, 8, 10
Parminder	5, 7, 8, 6, 5, 8, 6, 9

a   Calculate the mean score for James.
b   Calculate the mean score for Parminder.
c   The person with the highest mean was chosen to represent the class. Explain why this might not be the better person.

3   Two brands of carrot seeds were compared. The carrot harvest, in kilograms, from 10 packets of each brand are shown below.

Studley Seeds	12, 7, 11, 5, 13, 8, 7, 7, 11, 10
Super Seeds	9, 11, 8, 8, 9, 9, 11, 12, 8, 9

Which brand of seed do you think is better? Explain your answer.

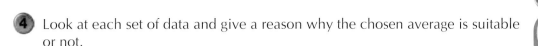

4 Look at each set of data and give a reason why the chosen average is suitable or not.

    a   4, 4, 5, 7, 7, 25   Mean
    b   3, 3, 3, 7, 9, 12, 12   Median
    c   5, 7, 7, 8, 9, 10   Mean
    d   9, 10, 12, 15, 18, 20   Mode

Practice

15C Statistical surveys

Write your own statistical report on one or more of the following problems.

Remember:
- Write down any hypotheses for the problem.
- Decide on your sample size.
- Decide whether you need to use a data collection sheet or a questionnaire.
- Find any relevant averages.
- Illustrate your report with suitable diagrams or graphs, and explain why you have used them.
- Write a short conclusion based on all the evidence.

The data can be collected from people in your class or year group, but it may be possible to collect the data from other sources outside school, such as friends and family.

1 Most people believe that eating organic vegetables will make you healthier.

(Hint: Design a questionnaire.)

2 On average, people can remember a number with seven digits after seeing it once.

(Hint: Design an experiment, including a data collection sheet.)

3 The average number of goals scored in a football match is fewer than 2.

(Hint: Use secondary sources such as newspapers and football annuals.)

Practice

15D Comparing experimental and theoretical probability

1 **a** **i** Write the letters S, P, I, N, N, E, R, S on eight pieces of paper and place them in a circle. Place a bottle in the middle.

ii   Copy this tally chart.

Letter	Tally	Frequency	Experimental probability	Theoretical probability
S				
P				
I				
N				
E				
R				

      iii   Spin the bottle 40 times and record your results in the tally chart. If the bottle stops exactly halfway between two letters, ignore the result and spin again.

      iv   Calculate the experimental probability for each letter. Write your answers as decimals.

      v   Calculate the theoretical probability for each letter. Write your answers as decimals.

b   In a game, if the bottle lands on an S or N you win, otherwise you lose.

      i   Calculate the experimental probability of winning.

      ii   Calculate the theoretical probability of winning.

      iii   Is the game fair? Explain your answer.

2   **a**   Two fair dice are thrown. The difference in their scores is calculated. For example, if 5 and 3 are thrown, the difference is 5 – 3 = 2. Copy and complete the table of differences below.

b   What is the theoretical probability of getting a difference of 3?

c   **i**   Throw two of your own dice 50 times. (If you do not have two dice, throw one twice.)

      ii   Record the differences in a tally chart.

      iii   Calculate the experimental probability of getting a difference of 3.

      iv   Do you think your dice are fair?

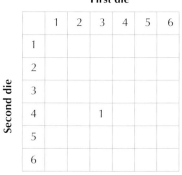

First die

	1	2	3	4	5	6
1						
2						
3						
4			1			
5						
6						

Second die

CHAPTER **16** Number **5**

Practice

16A   Multiplying decimals

Do not use a calculator to answer these questions.

1   Calculate each of the following.

   a 0.7×0.3 **b** 0.06×0.8 **c** 0.2×0.04 **d** 0.09×0.06

   e 0.5×0.4 **f** 0.03×0.08 **g** 0.6×0.6 **h** 0.4×0.008

2 Calculate each of the following.
  a 200 × 0.3 **b** 0.07 × 40 **c** 600 × 0.008
  d 0.04 × 9000 **e** 300 × 0.008 **f** 0.0007 × 300
  g 60 × 0.004 **h** 0.0009 × 300

3 Calculate each of the following.
  a 20 × 300 × 0.04 **b** 0.3 × 0.6 × 500 **c** 0.3 × 0.08 × 3000

4 A postage stamp weighs 0.04 g.
  a How much does a sheet of 300 stamps weigh?
  b How much does a box of 250 sheets weigh?

Practice

16B Dividing decimals

Do not use a calculator to answer these questions.

1 Calculate each of the following.

  a 0.8 ÷ 0.4 **b** 0.4 ÷ 0.02 **c** 0.06 ÷ 0.03
  d 0.9 ÷ 0.01 **e** 0.14 ÷ 0.07 **f** 0.36 ÷ 0.9
  g 0.2 ÷ 0.002 **h** 0.25 ÷ 0.05

2 Calculate each of the following.

  a 80 ÷ 0.2 **b** 300 ÷ 0.6 **c** 60 ÷ 0.04 **d** 400 ÷ 0.4
  e 50 ÷ 0.02 **f** 200 ÷ 0.05 **g** 32 ÷ 0.08 **h** 27 ÷ 0.3

3 Calculate each of the following.

  a 28 ÷ 700 **b** 1.8 ÷ 30 **c** 3.6 ÷ 200 **d** 0.3 ÷ 600
  e 60 ÷ 3000 **f** 0.24 ÷ 600

4 **a** A pack of 200 paper clips weighs 60 g. How much does one paper clip weigh?
  b A pack of 5000 staples weighs 0.12 kg. How much does one staple weigh?

Practice

16C Using a calculator

1 Use the memory keys on your calculator to work out each of the following. Write down any values that you store in the memory.

  a $\dfrac{15.8 + 17.1}{19.2 - 9.8}$

  b $\dfrac{92.51 - 26.75}{32 \times 0.15}$

  c 18.93 − (51.14 − 36.82)

  d $1.42 + \dfrac{1.5}{0.54 + 1.86}$

2 Calculate the answers to question 1 again, this time using bracket keys. Write out the key presses for each calculation. Which method uses fewer key presses?

3 Work out the value of each of these using the sign change key to enter the first negative number.

a −5 + 7 − 8
b −2.4 + 1.6
c −17 + 7 + 17 − 7
d −132 − 46 − 105

4 Use the square root key to work out each of the following.

a √1225
b √60
c √0.6
d √0

5 a Do you need to press ⊠ when multiplying a bracket on your calculator?

Try 5 (3 + 4) =

b Can you leave off the last bracket of a calculation?

Try 5 × (3 + 4 =

Practice

16D Fractions of quantities

1 Calculate each of the following.

a $\frac{2}{5}$ of £20
b $\frac{4}{7}$ of 56 kg
c $\frac{9}{10}$ of 110 ml
d $1\frac{3}{4}$ of 16 eggs
e $\frac{5}{8}$ of 400 g
f $2\frac{2}{3}$ of 60 miles
g $\frac{4}{11}$ of 121 seconds
h $\frac{19}{20}$ of 1000 tonnes

2 Calculate each of the following. Cancel your answers where appropriate and write them as mixed numbers.

a $4 \times \frac{3}{7}$
b $3 \times 1\frac{2}{5}$
c $7 \times \frac{3}{10}$
d $9 \times 2\frac{1}{6}$
e $\frac{7}{8} \times 6$
f $2 \times \frac{13}{14}$
g $5 \times 3\frac{7}{10}$
h $\frac{9}{16} \times 18$

3 $\frac{4}{5}$ of the annual rainfall in Quondu occurs in June. If the annual rainfall is 260 mm, what is the rainfall in June?

4 $\frac{2}{7}$ of the questions on a maths paper are arithmetic. There are 42 questions altogether. How many are arithmetic?

 5 $1\frac{5}{9}$ cans of paint are needed to paint a fence. How many cans are needed to paint six identical fences?

Practice

16E Percentages of quantities

1 Calculate each of the following.

a 80% of £2.40
b 65% of 80 kg
c 5% of £3200
d 44% of 600 people

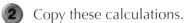

2 Copy these calculations.
   a   … % of 40 = 18
   b   … % of 300 = 96
   c   … % of 250 = 100

The missing percentages are 32%, 40% and 45%.
Complete each calculation with the correct percentage.

3 A bag contains 200 chocolate beans. Anna eats 20% and then gives the bag to Raj. Raj eats 35% of the remaining beans. Siobhan gets what is left over.
   a   How many beans does Anna eat?
   b   How many beans does she give to Raj?
   c   How many beans does Raj eat?
   d   How many beans does Siobhan get?
   e   What percentage of the original bag does Siobhan get?

4 Copy and complete this table.

	a	b	c	d	e
Decimal	0.8			0.875	
Fraction			$\frac{7}{20}$	$\frac{19}{25}$	
Percentage		6%			

CHAPTER **17** Algebra **5**

Practice

17A Solving equations

1 Solve each of the following equations.
   a $2n + 7 = 25$ b $8k + 20 = 68$ c $11f + 13 = 101$
   d $9g + 55 = 370$

2 Solve each of the following equations.
   a $3w - 8 = 25$ b $5e - 30 = 10$ c $12m - 4 = 92$
   d $6v + 4 = 28$ e $9j - 17 = 55$ f $30x + 100 = 700$

3 Solve each of the following equations.
   a $2(2x - 1) = 10$ b $2(5x + 3) = 26$ c $3(4x - 1) = 45$
   d $2(2x - 5) = 44$ e $5(4x + 3) = 80$ f $3(2x + 9) = 105$

4 Solve each of the following equations.
   a $5x + 2 = 3x + 12$ b $6x - 1 = 2x + 7$ c $3x + 4 = 9 - 2x$
   d $4x - 5 = 2x + 2$ e $10x + 1 = 4 - 5x$

Practice

17B Formulae

1 This formula is used to generate odd numbers: $m = 2n + 1$. Find m when:

a $n = 4$ **b** $n = 16$

 2 This formula can be used to convert pounds (£) into euros (€):

$$E = \frac{8P}{5}$$

where P is the number of pounds and E is the number of euros.

a Use the formula to convert £40 to euros.

b Use the formula to convert £18 to euros.

 3 The time needed to roast a joint of meat is given by the formula

$$T = 30w + E$$

where T is the cooking time in minutes, w is the weight of the joint in kg and E is extra time.

a Calculate the cooking time for a joint weighing 2.5 kg which needs 25 minutes of extra time.

b Calculate the cooking time when $w = 1.8$ kg and $E = 15$ minutes.

4 This metal plate was made by cutting a square hole from a square piece of metal. Its perimeter is given by the formula

$$P = 4(D + d)$$

where P is the perimeter, D is the length of the outer side and d is the length of the inner side.

a Calculate the perimeter when the outer side is 25 mm and the inner side is 18 mm.

b Calculate the perimeter when $D = 4.6$ cm and $d = 1.25$ cm.

Practice

17C A square investigation

Each shape is made using an inner square and an outer square. The side of the outer square is shown below each shape.

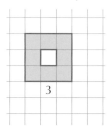

3

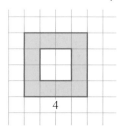

4

1 Copy this table.

Side of outer square, s		3	4	5	6	7	8
Number of shaded squares, n							

2 Count the numbers of shaded squares and write them in your table.

3 Draw more shapes to complete the table. The inner square must be as large as possible.

4 What do you notice about the numbers of shaded squares in the table?

5 Write down a rule that gives the number of shaded squares, n, if you are given the side, s. (Hint: Start by multiplying by 4.)

6 Write your rule using algebra.

Practice **17D Graphs from the real world**

 1 The graph shows how long it takes for a computer printer to print a number of photos. Use the conversion graph to answer the following questions.

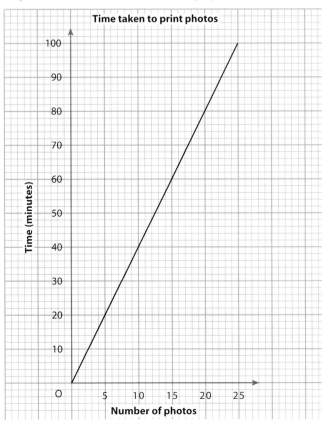

a How long does it take to print   **i** 10 photos   **ii** 22 photos?

b How many photos can be printed in   **i** 80 minutes   **ii** 48 minutes?

FM **2** A clothing mail order company charges £2.50 per item for postage and packing.

a Copy and complete the table.

Number of items	2	4	6	8	10	12	14
Postage and packing (£)			15				

b Draw a graph using these scales:
    x-axis (Number of items) 2 cm to 2 items
    y-axis (Postage and packing) 2 cm to £5
Use your graph to answer the following questions.

c What is the postage and packing charge for   **i** 6 items   **ii** 13 items?

d How many items can be delivered for a postage and packing charge of
    i £10   **ii** £27.50?

FM **3** The cover of a hardback novel weighs 50 g. Twenty pages weigh 40 g.

a Copy and complete the table.

Number of pages	100	200	300	400
Weight of book (g)			650	

b Draw a graph using these scales:
    x-axis (Number of pages) 2 cm to 100 pages
    y-axis (Weight of book) 2 cm to 100 g
Use your graph to answer the following questions.

c What is the weight of a novel with   **i** 150 pages   **ii** 320 pages?

d How many pages does a novel have that weighs   **i** 730 g   **ii** 590 g?

Write an equation for each question. Then solve your equation.

Practice

17E Using algebra to solve problems

Example

Three children and two adults each buy a train ticket. An adult ticket costs £3.50 more than a child ticket. The total cost of the tickets is £32. Find the cost of a child ticket.

Working

Let x be the cost of a child ticket.

An adult ticket costs $x + 350$

3 child tickets + 2 adult tickets cost 3200p altogether

Writing this as an equation gives:

$$3x + 2(x + 350) = 3200$$

$$3x + 2x + 700 = 3200$$

$$5x + 700 = 3200$$

$$5x = 3200 - 700$$

$$5x = 2500$$

$$x = 500$$

Answer   A child ticket costs £5. So an adult ticket costs
£5 + £3.50 = £8.50.

Check   3 × £5 + 2 × £8.50 = £15 + £17 = £32

1 Large eggs weigh 8 g more than medium eggs. A large egg and a medium
egg have a total weight of 72 g. Let the weight of a medium egg be x grams.

    a   Write down in terms of x the weight of a large egg.
    b   Set up and simplify an equation in x using the above information.
    c   Solve the equation to find the weight of a medium egg.

2 Barry thinks of a number, multiplies it by 3 and then subtracts 8. His answer
is 28. Let x be the number that Barry thought of.

    a   Write down an equation in x to represent the above problem.
    b   Solve the equation to find the number that Barry thought of.

3 The sum of three consecutive numbers is 75. Let the middle number be x.

    a   Write down, in terms of x, the other two numbers.
    b   Set up and simplify an equation in x to show the above information.
    c   Find the values of the three numbers.

4 On Monday a taxi driver drove x miles. On Tuesday she drove 20 miles
more than Monday and on Wednesday she drove four times the distance
she drove on Monday.
In total she drove 350 miles over the three days.
    a   Write down in terms of x how many miles she drove on Tuesday.
    b   Write down in terms of x how many miles she drove on Wednesday.
    c   Set up an equation in x to represent the above information.
    d   How many miles did she drive on Monday?

5 A pen costs 7p more than a pencil. Hannah buys six pens and four pencils
for £1.02. What is the cost of a pencil?

6 £6.80 of Bingo prize money was shared between three children and two
adults. Each adult received 40p more than each child. How much did each
child receive? (Hint: Work in pence.)

7 The total cost of 10 pillows and 15 pillow cases is £175. A pillow costs £6.50 more than a pillow case. How much does a pillow case cost? (Hint: Work in pounds.)

CHAPTER 18 Geometry and Measures 5

Practice

18A Polygons

1 a b c  d e

    i   Name each polygon, e.g. heptagon.
    ii Which polygons are regular?
    iii Describe each polygon as concave or convex.

2 Draw each of these polygons.

    a   A pentagon with one right-angle and one reflex angle.
    b   A hexagon with rotational symmetry, order 2.
    c   An octagon with exactly two lines of symmetry.
    d   A hexagon with one line of symmetry and two reflex angles.

3 Use squared paper to draw eight different kinds of polygon inside a 3 by 3 grid. Use a different grid size for each polygon. Describe each polygon, e.g. convex quadrilateral.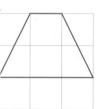

Practice

18B Tessellations

Make a tessellation from each of the following shapes, if possible. Use a square grid to help.

a   **b** 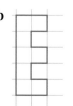   **c** 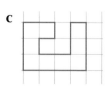   **d**

1 Which of the following are nets for the half cylinder shown?

a   b   c   d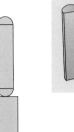

2 Stick a photocopy or a tracing of this net onto a sheet of thin card. Then cut out the net. Next, fold and glue it to make an icosahedron.

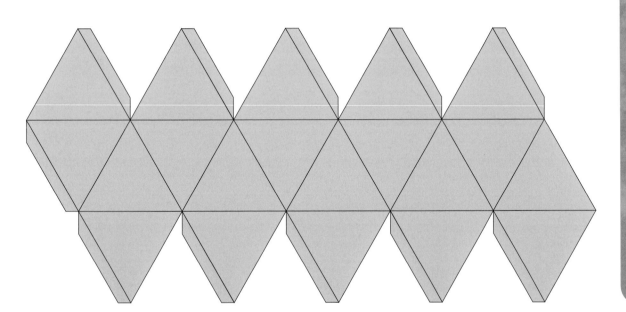